~Famous Inventors~

Thomas Edison

Inventor

Carin T. Ford

Enslow Publishers, Inc.
40 Industrial Road
Box 398
Berkeley Heights, NJ 07922
USA

PO Box 38
Aldershot
Hants GU12 6BP
UK

http://www.enslow.com

Library of Congress Cataloging-in-Publication Data

Ford, Carin T.
Thomas Edison : inventor / Carin T. Ford.
p. cm. — (Famous inventors)
Summary: Profiles a man whose childhood interest in experimentation led him to create more than one thousand inventions, including the phonograph, moving pictures, and a storage battery.
Includes bibliographical references and index.
ISBN 0-7660-1860-1 (hardcover)
1. Edison, Thomas A. (Thomas Alva), 1847–1931—Juvenile literature. 2. Inventors—United States—Biography—Juvenile literature. 3. Electric engineers—United States—Biography—Juvenile literature. [1.Edison, Thomas A. (Thomas Alva), 1847–1931. 2. Inventors.] I. Title. II. Series.
TK140.E3 F64 2002
621.3'092—dc21

2002000488

Printed in the United States of America

10 9 8 7 6 5 4 3 2 1

Illustration Credits: Library of Congress, pp. 3, 6(L), 6(R), 11, 14–15, 16(R), 17, 19, 25, 26–27, 28; Photo courtesy of Western Union Financial Services, Inc., p. 10; Schenectady Museum Archives, pp. 4, 23; Simon Rosenstein, pp. 1, 2, 21, 22, 27, 29T, 29B, 30; U.S. Department of the Interior, National Park Service, Edison National Historic Site, pp. 8, 16(L), 20.

Cover Credits: Edison photograph courtesy of Library of Congress; lightbulb photo, Simon Rosenstein.

Table of Contents

1 What a Curious Boy! 5

2 Going to Work 9

3 Early Inventions 13

4 The Wizard 18

5 Moving On 24

Timeline 29

Words to Know 30

Learn More About Thomas Edison (Books and Internet Addresses) 31

Index 32

Thomas Alva Edison

Chapter 1

What a Curious Boy!

Young Thomas Alva Edison liked to experiment. He liked to mix chemicals and see what would happen.

One day, Al—as he was called—and some of his friends were trying experiments in an old office building. They mixed a few chemicals together. Suddenly, there was an explosion. A corner of the building was ruined. Al and his friends were burned.

Al was not badly hurt—and he did not stop

Al was born in this house. As a boy, he asked his mother, left, many questions.

experimenting. During his life, he would create more than one thousand inventions. Many, like his improved lightbulb, changed people's lives.

Al was born on February 11, 1847, in Milan, Ohio. His parents, Samuel and Nancy Edison, had seven children. Al was the youngest. He was born with a large head, and the doctor worried that something might be wrong with his brain. Samuel and Nancy Edison kept a close eye on their son.

Al was an active and very curious little boy. He asked questions all day long. If his parents could not give him an answer, Al would say, "Why *don't* you know?"

When Al was seven, his family moved to Port Huron, Michigan. There he was sent to school for the first time. But it did not last long.

Al was not a good student. He did not pay attention in class. He drew pictures in his notebook. He asked the teachers questions, instead of coming up with the answers.

The teachers said Al was "addled," or confused. When Al's mother heard this, she became angry. Al's mother used to be a teacher, and she decided to teach Al herself.

Al's mother gave him many books, and he quickly learned to read. He enjoyed reading about history and religion. But Al's favorite books were about science.

Al's early experiments did not always work. But he never stopped trying new things.

Chapter 2

Going to Work

Chemistry books were the most interesting to Al. It was not long before he had filled his basement with many different bottles of chemicals. He labeled all of them "poison" so no one would touch them. His mother worried that he would blow up the house.

Al turned the basement into a laboratory and did many experiments. He even built his own telegraph machine. A telegraph uses electric pulses to send messages over a wire. The messages sound like clicks. Short and long clicks—called dots and dashes—stand for letters and numbers. Al set up a wire between his house and his friend's house. The two boys sent each other messages on Al's telegraph.

When he was twelve, Al decided to get a job. He sold snacks and newspapers on a train that traveled between Port Huron and Detroit, Michigan. For eight hours each day, he walked up and down the aisles of the train selling papers, apples, peanuts, molasses, and sandwiches.

Al used a telegraph machine like this one.

Every day the train stopped in Detroit for several hours. Al made good use of his free time. He set up a laboratory in the baggage car to do experiments. But one day, he accidentally started a fire

Al earned money selling newspapers and snacks to travelers on the railroad.

with his chemicals. The conductor told Al he must stop doing experiments.

Always busy, Al started his own newspaper when he was fifteen years old. He wrote the stories and printed the paper while he was on the train.

It was during this time that Al began to lose his hearing. No one is sure why this happened. Al said he was late for the train one day and ran to catch it. When the conductor lifted Al aboard by his ears, he felt something snap inside his head. He said that his deafness started not long after this.

But Al also had many ear infections as a child. In the mid-1800s, there was no medicine to treat this. It is possible that Al's ear infections harmed his hearing.

Over time, Al's hearing became even worse. But he said that he found being deaf useful. His near-silent world allowed him to focus better on his work.

And Al could not seem to stop working. Throughout his life, Al slept only four or five hours a night. During the day, he would take short naps. He could fall asleep anywhere for a few minutes—on a desk, under a table, or sitting in a chair. Al did not want to waste time sleeping. There was too much he wanted to do.

Chapter 3

Early Inventions

Telegraph machines continued to interest Al. But he still did not know how to use one properly.

One day, fifteen-year-old Al saved a little boy's life. Al saw a three-year-old playing on the railroad tracks. Al pulled the child to safety before a train could hit him. The boy's father was a telegraph operator. He rewarded Al by teaching him how to use the telegraph.

Telegraph operators were needed all over the country. Now Al could get a job. For the next few years, Al traveled from job to job in cities across the United States and even Canada. But he realized that he wanted to do more than work telegraphs. He wanted to improve them. Soon Al got his chance.

Al moved to Boston when he was twenty-one years old. He spent all his time with his inventions. He enjoyed juggling several projects at once. Besides trying to make a better telegraph, he also worked on

fire alarms and machines to send handwritten messages.

In 1868, Al invented a vote-recording machine for some lawmakers in Washington, D.C. They could vote just by pushing a button. But the lawmakers did not want to buy Al's machine. They preferred to say their votes out loud, so others would know how they voted. Still, Al's invention was the first step toward today's voting machines.

Al then made a promise to himself. He would not waste his time creating inventions that people did not want. From now on, he would only invent things that people were interested in buying.

Telegraph operators in Western Union offices were busy all the time.

Al moved to New York City the next year, 1869. He had little money and had to beg for food. He slept in an office building. One day, a machine in the office stopped working and no one could fix it. Al took a look at it, and he had it working in two hours. Instantly, the company hired him.

Soon Al opened a small shop in New Jersey. He worked on many electrical devices. For example, he created a new kind of telegraph that could send four messages at the same time, instead of just one.

As always, Al spent most hours of the day and night working. He believed he could do anything if he worked hard and used common sense.

Al married Mary Stilwell on Christmas Day in 1871.

The wedding could not begin until Al finished his experiment.

In 1871, when he was twenty-four, Al married Mary Stilwell. She was sixteen years old and worked for his company. In the next few years, they had three children: Marion, Thomas Alva Jr., and William.

Al now had enough money to do what he had always wanted. In 1876, when he was almost thirty, he hired a few men and set up a big laboratory in Menlo Park, New Jersey. This was the first time anyone had gone into the business of making inventions.

Before long, Al was known as the "Wizard of Menlo Park."

Chapter 4

The Wizard

Almost all of Al's time was spent working on inventions. He would think about what people needed—and then invent it.

In 1876, another inventor, Alexander Graham Bell, stunned the world with the invention of the telephone. But there were two problems with Bell's telephone. It worked only over short distances—less than twenty miles. And even then, a person had to shout into it to be heard.

Al began to experiment. He wanted to make a telephone that worked better. He soon found that if he used a chemical element called carbon in the telephone, the voices could be heard clearly and could travel over long distances.

THE EDISON PHONOGRAPH
TRADE
Thomas A Edison
MARK
HAVE YOU HEARD IT?

While working on the telephone, Al got an idea for another invention. What if sounds could be saved so they could be played back and listened to later?

Again Al began to experiment.

He created the phonograph in 1877. People called it the talking machine. To record sounds, a needle made grooves on tin foil. Al then used a different needle to play back the sounds.

He tested the machine by reciting "Mary Had a Little Lamb" into it. When the message played back, Al said he was never so surprised in his life!

Al became well known because of the phonograph. This invention would always be his favorite. He called it his "baby."

Al now turned his attention to electricity. In the 1870s, people used candles, oil lamps, or gas to light

their homes. Electric lights had already been invented. But their light was much too bright and hot to use in the home. Al knew a light was needed that would produce a softer glow.

Once again, Al began to experiment.

At first, people could not use the lightbulb because most homes did not have electricity.

He developed an electric lightbulb similar to the one we use today. It gave off a soft light. And it lasted a long time without burning out.

Along with this invention, Al also created a complete lighting system. Until this time, people did not have light switches or sockets in their homes. There was no way they could use electricity. But Al came up with a way to deliver electric light to homes and businesses. Now everyone could use the lightbulb.

By the time he was in his thirties, Al had become one of the most famous men in the world.

Edison and his assistant, Francis Jehl, test an electric lightbulb.

Chapter 5

Moving On

The laboratory was known as the invention factory. There, Al made many wonderful inventions. But after his wife died in 1884, Al did not spend much time at Menlo Park. It made him too sad.

Two years later, he married Mina Miller. They moved to West Orange, New Jersey. In the next several years, they had three children: Madeleine, Charles, and Theodore.

Al built a new laboratory much larger than the one in Menlo Park. He put in the best machines. He made sure he had all the books and science journals he might need.

Mina Miller, a stylish young woman, became Edison's second wife in 1886.

Al had always been interested in photography. He began to wonder if he could create an invention that could record photos and then play them back.

Al also began working on movies. He developed a movie camera and a projector. He also built the world's first movie studio where films could be made. Covered with black tar paper, the building was called Black Maria.

At age forty-seven, Al began to make movies. Many famous actors traveled to his New Jersey

Edison invented a machine that flipped photos one after the next. One of his first movies showed a man sneezing.

studio to perform there. Some movies showed boxing matches. Others showed ballet dancers and acrobats.

Al was getting older, but there was still so much he wanted to do.

He went on to invent a battery that could be used over again. Al also worked for the U.S. government during World War I, developing devices to save American ships from torpedoes. He was proud that he never invented a weapon that could kill anyone.

As Al got older, he also experimented with his diet. For a long time, he had been eating only very

small amounts at every meal. He finally decided to stop eating any food at all. For a while, Al lived on nothing but several glasses of milk a day.

Thomas Alva Edison died on October 18, 1931. He was eighty-four years old. On the day he was buried, Americans turned off their electric lights for one minute to honor him.

Al never thought there was anything special about what he did. He always said that all an inventor needed was an imagination and a pile of junk.

Al invented the first talking doll in 1886.

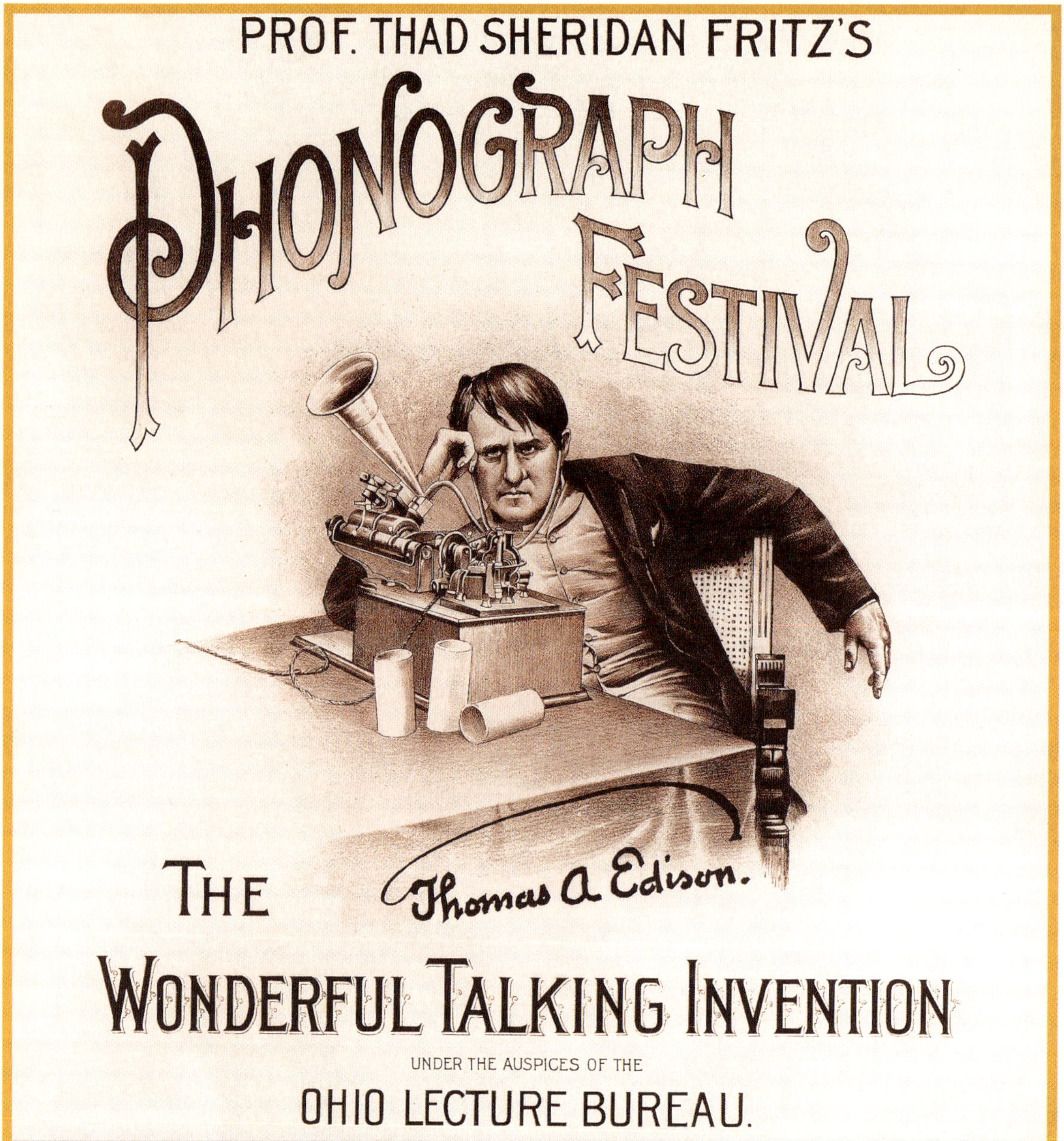

People were very excited to hear about Edison's "wonderful talking machine."

Timeline

1847~Born in Milan, Ohio.

1859~Works on Grand Trunk Railway.

1863~Works as telegraph operator.

1876~Opens Menlo Park, New Jersey, laboratory.

1877~Invents phonograph.

1879~Invents first practical lightbulb.

1886~Builds new lab in West Orange, New Jersey.

1891~Invents movie camera.

1912~Develops "talking" movies.

1917~Works on defense for the U.S. government during World War I.

1931~Dies October 18 at age eighty-four.

Words to Know

chemistry—A branch of science that deals with what things are made of.

electricity—A kind of current that provides the power to run lights and different kinds of machines.

experiment—To test or try something.

laboratory—A place where science experiments can be done.

phonograph—An instrument that is able to save sounds and play them back.

projector—A machine for showing moving pictures on a screen.

telegraph—A machine that uses electric pulses to send messages over a wire.

Learn More

Books

David A. Adler, *Thomas Alva Edison: Great Inventor* (New York: Holiday House, 1990).

Barbara Mitchell, *The Wizard of Sound: A Story about Thomas Edison* (Minneapolis: Carolrhoda Books, 1991).

Pam Robson, *Electricity* (New York: Aladdin Books, 1992).

Internet Addresses

The Edison National Historic Site has lots of great information
<http://www.nps.gov/edis/home.htm>

Thomas A. Edison and the Menlo Park Laboratory
<http://www.hfmgv.org/exhibits/edison/>

Index

B

Bell, Alexander Graham, 18
Black Maria, 25
Boston, Massachusetts, 14

D

Detroit, Michigan, 10

E

Edison, Charles (son), 24
Edison, Madeleine (daughter), 24
Edison, Marion (daughter), 17
Edison, Mary Stilwell (wife), 16, 17, 24
Edison, Mina Miller (wife), 24, 25
Edison, Nancy (mother), 6, 7, 8, 9
Edison, Samuel (father), 6, 7
Edison, Theodore (son), 24
Edison, Thomas Alva
- childhood, 5–8, 9–10
- death, 27
- education, 7, 8
- hearing problems, 11–12
- jobs, 10, 14
- marriages, 17, 24

Edison, Thomas Alva Jr. (son), 17
Edison, William (son), 17
electric lightbulb, 22, 23
electricity, 21–22

J

Jehl, Francis, 23

M

Menlo Park laboratory, 17, 24, 25
Menlo Park, New Jersey, 17
Milan, Ohio, 6
motion pictures, 25–26
movie camera, 25

N

New York City, 16

P

phonograph, 20, 21, 28
Port Huron, Michigan, 7, 10

T

telegraph, 9–10, 13, 14, 16
telephone, 18, 19

V

vote-recording machine, 15

W

Western Union, 15
West Orange laboratory, 25
West Orange, New Jersey, 24
"Wizard of Menlo Park," 17
World War I, 26